INSIDE SCIENCE

VOYAGE

Adrienne M. Frater

CONTENTS

COME ABOARD

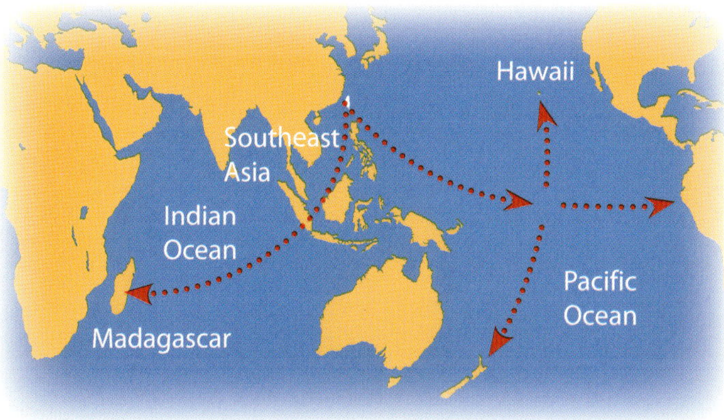

An amazing thing happened 3,500 years ago. Some Austronesian people from Southeast Asia sailed across the Indian Ocean to Madagascar. Others sailed out into the **vast** Pacific Ocean. No one had ever sailed so far before. At last, human beings were exploring the largest ocean on earth. Their **descendants** would eventually discover Hawaii.

Now scientists explore the world's oceans in marine research vessels. The science and technology they use is a mixture of the modern and the very old.

An early Hawaiian voyaging canoe

Come aboard a modern research vessel. We're heading off on a **voyage** of discovery.

"Vessel" is another word for a ship or a boat. What's the difference between a ship and a boat? Here's a clue. You find boats on ships, but you don't find ships on boats. This definition works for everything except submarines. They're always called boats!

Even if you've been on board a ship before, you may have a few questions.

WHY *DO* SHIPS FLOAT?

Before a ship like this goes to sea, it is loaded with everything the crew needs. This includes food, water, scientific equipment, and the crew's personal gear.

The crew

A submersible

Already on board are the **navigation** and safety equipment and a submersible— a small submarine to use for deep-sea exploration. It's quite a load.

So why doesn't this ship sink? All this stuff weighs a lot!

One of the reasons a heavy ship can stay afloat is a force called upthrust. When you get in a bathtub, the water rises. In the same way, a ship pushes water out of the way. The water also pushes back on the ship's **hull**. This push is called upthrust. A ship floats as long as the weight of the ship is less than the upthrust.

If you look at the side of a ship, you'll see marks painted on the hull. They're load lines. The crew knows that once the water reaches the line, it's not safe to load any more **cargo** on board.

Find the Load Line

To see how upthrust works, try this experiment.

You will need:

- a sink or a tub of water
- a model boat
- gravel
- a pen

What to do:

1. Fill the sink or tub with water and put the boat in the water.
2. Start filling the boat with gravel.
3. Watch what happens. How much gravel can the boat hold before it sinks?
4. Find where the load line should go on your model and mark it.

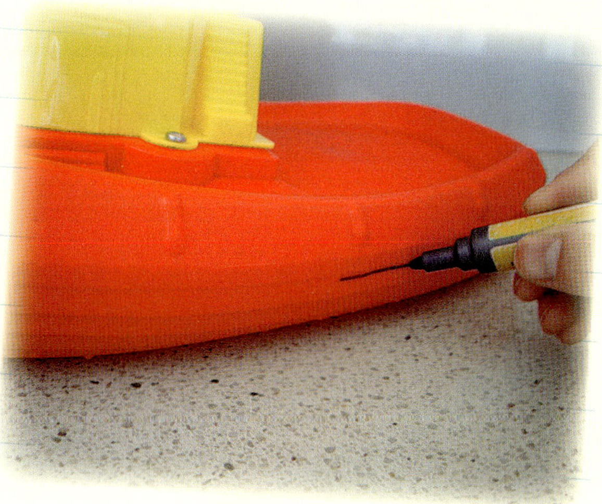

Tell a friend:

Invite a friend to load your model boat. Show them how the load line works.

A Plimsoll line

Samuel Plimsoll's Unseaworthy Shipping Bill (1876) was a law that forced ship owners to paint load lines on ships' hulls. They are now called Plimsoll lines in honor of him.

Stability

On calm days, it's easy to walk on the **deck** of a ship. On rough days, you'll need to be careful—you might fall down! If the ocean gets really rough, ships can tip over, too. There's no need to worry about this ship tipping over, though, because it's **stable**. Ensuring stability is extremely important when loading a ship. The heaviest objects are stored in the lowest part of the ship to keep the center of gravity as low as possible.

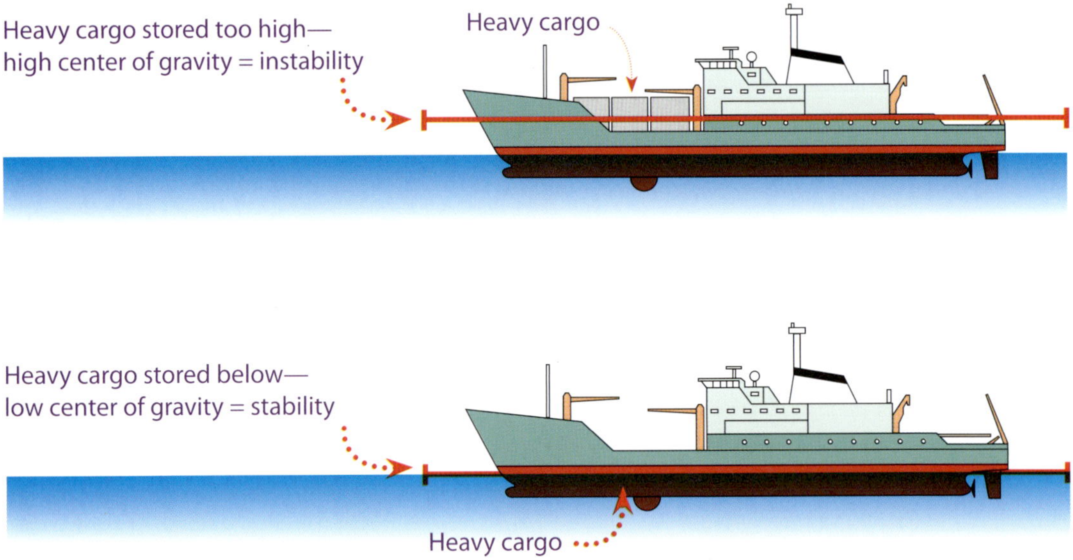

Everything is stored as evenly as possible, so that the ship isn't too heavy on one side. Anything that might move in a storm is tied down. **Watertight** doors are shut.

WHERE ARE WE?

Is all your gear safely stowed away? Good. It's time to check out your **berth**. Sleep well, but be warned that when you wake up tomorrow morning, you won't be able to see any land!

A berth

A porthole

How will we know where we are if we can't see land? People have traveled on the oceans for thousands of years. The problem with staying near the shore is that it's too dangerous—a ship can run onto rocks or reefs. Sailors have long known that it is safest to keep far out at sea. They don't need to be able to see land to know where they are.

Ancient sailors like the Austronesians used their knowledge of the sun, stars, seasons, waves, winds, currents, clouds, and birds to guide them. They also developed devices to help them find their way.

A mattang—a stick map used by ancient navigators in the Pacific

Later sailors developed other instruments.

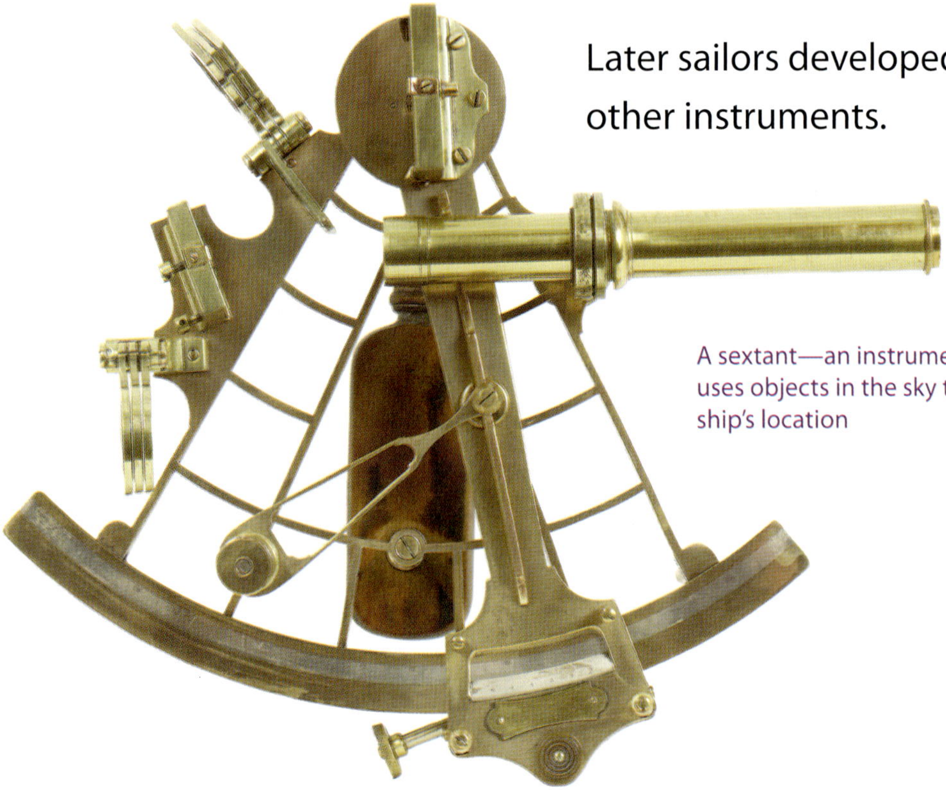

A sextant—an instrument that uses objects in the sky to find a ship's location

GPS

Today's navigators use electronic instruments. They look at computer screens using systems like the Global Positioning System, or GPS. A ship's GPS receiver measures its distance from satellites in orbit above the earth.

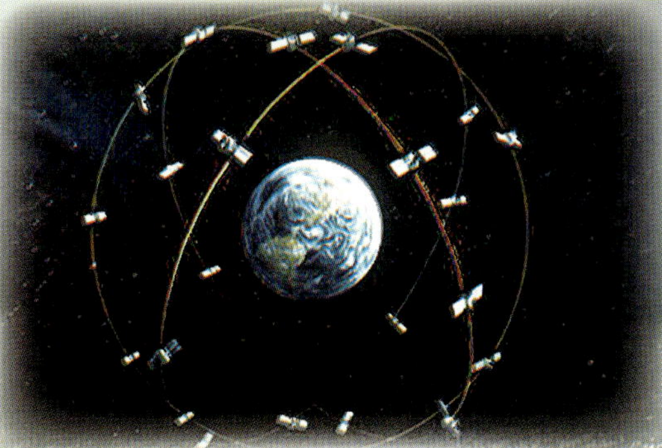

GPS satellite system

Close-up of satellite

Finding the ship's position

Radar and Sonar

When you are on a ship, it's good to know where you are. It's also good to know where other things are. You don't want to bump into them in the dark!

In 1912, the passenger ship Titanic *hit an iceberg and sank. More than 1,500 people died. After this tragedy, scientists started looking for a way to "see" icebergs in the dark.*

Radar (**ra**dio **d**etection **a**nd **r**anging) uses echoes to detect objects such as islands and other ships. Pulses of radio waves pass through the air. When they hit an object they bounce back as echoes. By timing the gaps between the pulses and the echoes, a computer can calculate how far away something is.

A ship's radar screen

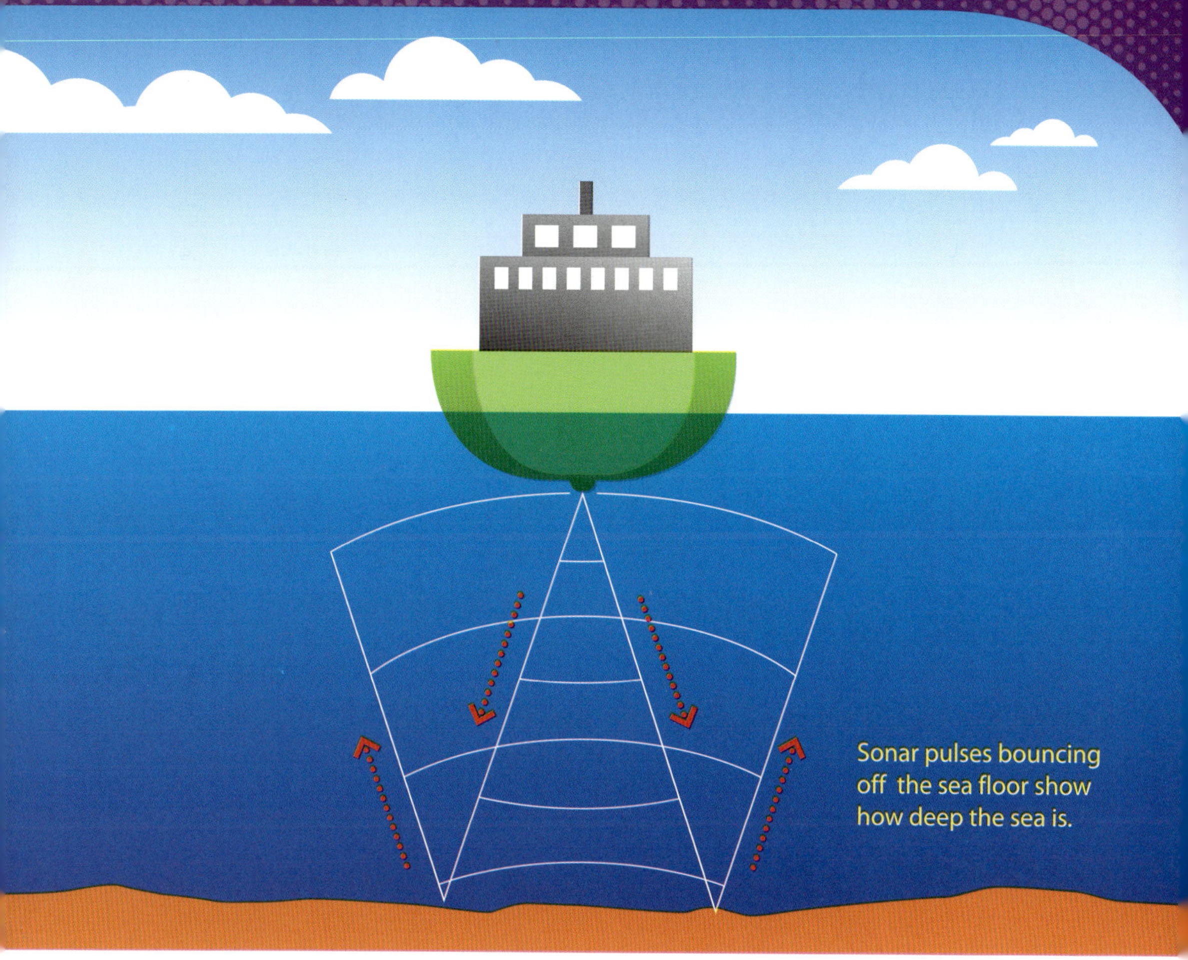

Sonar (**so**und **n**avigation **a**nd **r**anging) uses echoes, too, but with sonar, the pulses are made of sounds passing through water. They are reflected back by things under the water. Using sonar, you can tell how deep the sea is when the echoes bounce off the seafloor.

Of course, you can't rely on GPS, radar, and sonar alone. What if the equipment breaks down? In an emergency, navigators still need to know the science involved. They must be able to do some of the **calculations** themselves.

WEATHER AT SEA

Sailors also need to know about the weather. In fact, take a look outside. Notice that the waves are getting larger. See those dark clouds piling up? Hear the wind?

Do you know the difference between weather and climate? Is a storm coming? Is it starting to rain? That's *weather*. What's the weather like year after year? The answer to this question is *climate*.

Barometers measure changes in air pressure. Falling air pressure means that bad weather is coming.

Let's download a map from a weather satellite.

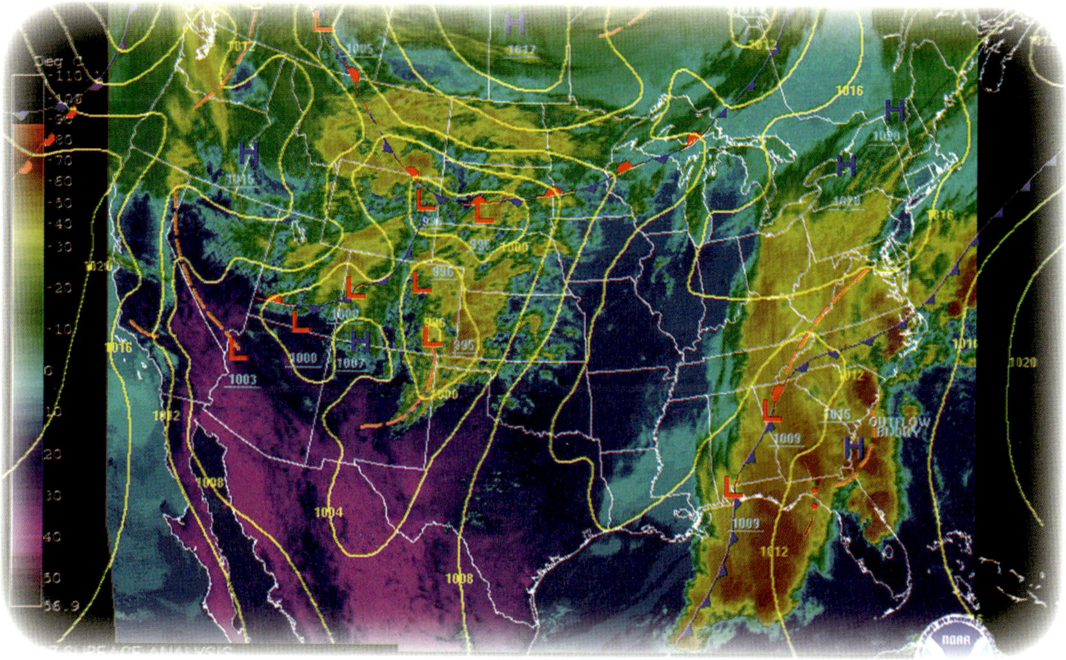

If the storm is going to be really severe, we'll alter course and try to avoid it. We'll use an anemometer to measure the wind's speed. It's dangerous out here in really bad weather.

An anemometer

Can you believe that the first anemometer was made in Italy in 1430? That's well over 500 years ago!

The Beaufort Scale was invented to help sailors estimate the force of the wind. It is still widely used both at sea and on land.

The Beaufort Wind Force Scale

Force	Strength	Miles per hour	Kilometers per hour	Effect at sea	Effect on land
0	Calm	0	0	Flat sea	Any smoke rises straight up
1	Light air	1–3	1–5	Rippled sea	Any smoke drifts
2	Light breeze	4–7	6–11	Small wavelets	Leaves rustle
3	Gentle breeze	8–12	12–19	Large wavelets	Twigs move
4	Moderate breeze	13–18	20–28	Small waves	Dust and paper blow around
5	Fresh breeze	19–24	29–38	Moderate waves	Small trees move
6	Strong breeze	25–31	39–49	Large waves	Umbrellas turn inside out
7	Near gale	32–38	50–61	Waves heap up, white spray	Hard to walk
8	Gale	39–46	62–74	Waves break	Twigs break off trees
9	Strong gale	47–54	75–88	Lots of spray	Branches blow down
10	Storm	55–63	89–102	Sea becomes white	Trees uproot
11	Violent storm	64–72	103–117	Exceptional waves	Buildings are damaged
12	Hurricane	Over 72	Over 118	Huge waves, difficult to see	Devastation

Become a Wind Estimator

You can estimate the wind's force just by looking around you!

You will need:

- the Beaufort Scale
- a notebook
- a pencil

What to do:

1. Put a copy of the Beaufort Scale in the front of your notebook.
2. For the next two or three weeks, at noon each day, in the same place if you can, look at the effect of the wind on the sea or the land.
3. Using the Beaufort Scale, estimate the wind's strength.
4. Each day, on a separate page, record the date, the place, and the wind's strength.

Impress your friends:

After a few weeks, you'll start to be very good at estimating the wind's strength and speed. Impress your friends by telling them how fast the wind is blowing.

SCIENCE AT SEA

Tracking the weather is not the only thing scientists do on research voyages. On this voyage, we're going to explore the depths of the Gulf of Mexico and track a shark. Remember the submersible we've got on board? We'll use it to explore deep underwater.

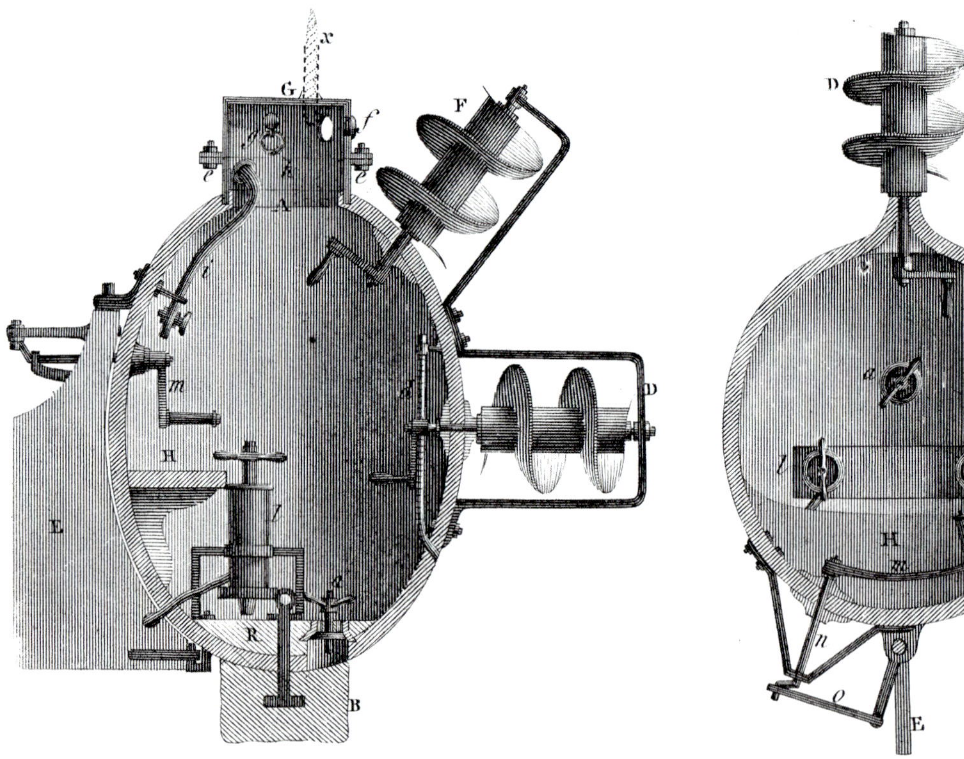

The inside of the *Turtle*

One of the first submersibles was the *Turtle*, designed by the American inventor David Bushnell in 1776.

Submersibles take scientists deep under the ocean's surface.

Did you know that the deepest place in the ocean is in the Marianas Trench, in the Pacific Ocean? It's nearly 7 miles (11 kilometers) below sea level. Mount Everest is only 5.5 miles (9 kilometers) high, by comparison.

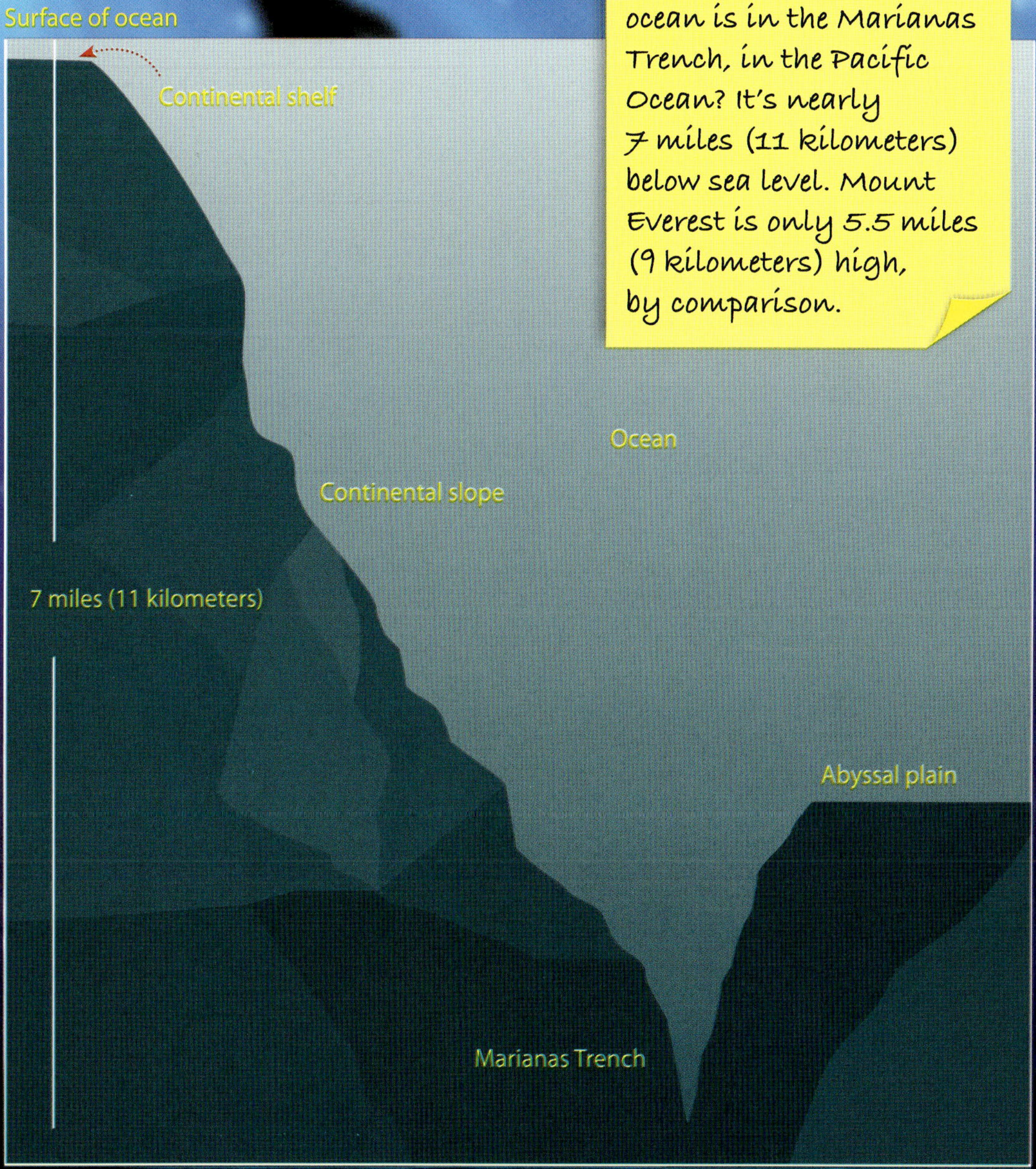

Imagine the thrill of studying things that no one has ever seen before!

It's so dangerous down there that we're going to use a remotely operated vehicle (ROV), a special kind of submersible.

In the Gulf of Mexico, an octopus explores the manipulator arm of an ROV.

Marine scientists also track the movements of animals such as whales, turtles, and sharks. They attach satellite tags to the animals so that they can follow their movements.

Did you know that you can follow the movements of tagged whales, turtles, and sharks on the Internet?

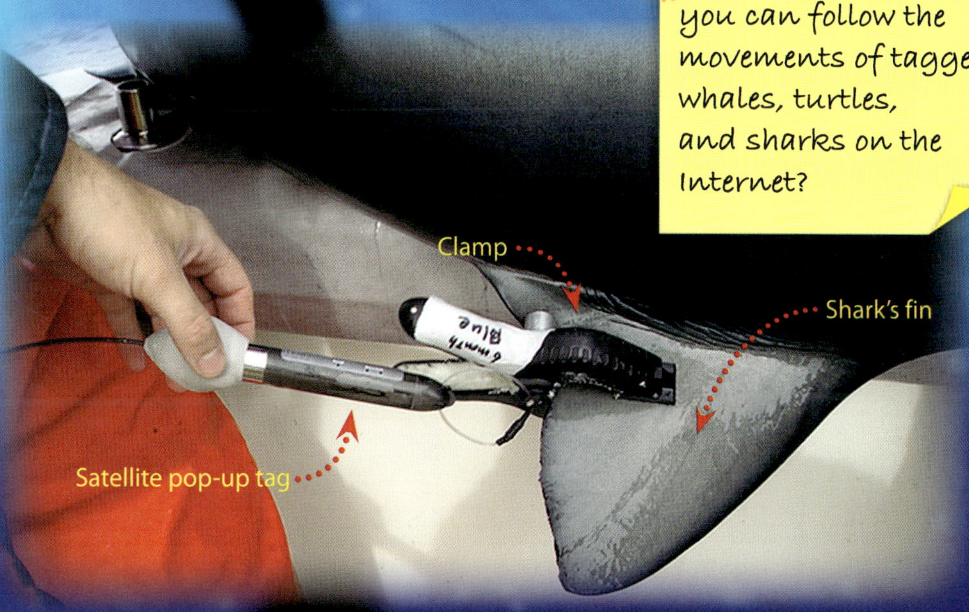

Tagging a shark using a radio tag. The signal is picked up by GPS satellites.

One tagged great white shark recently swam across the Indian Ocean from Australia to South Africa in 99 days. Then she turned around and swam right back! Before this, scientists had no idea that sharks could swim so far, so fast! The Indian Ocean is the third biggest ocean in the world and is 6,200 miles (9,977 kilometers) wide. The United States is only 2,600 miles (4,184 kilometers) wide.

WANT TO STAY ON BOARD?

The scientists who study the ocean and its plant and animal life use technology and science that has taken thousands of years to develop—but there is still much to learn! We know more about what it's like in some places on the moon than we know about many parts of the ocean.

This voyage is almost over, but if you'd like to continue the trip you started with this book, think about going to sea some day on a real research vessel. You'll have some incredible adventures. Good luck!

GLOSSARY

berth—where you sleep on a ship

calculations—the math you use to work out the answer to a problem

cargo—the things a ship carries

deck—the floor on a ship

descendants—later generations of related people

hull—the frame of a ship, excluding things above the ship's deck, such as masts

navigation—to do with locating where you are and finding your way

stable—not likely to tip over, balanced

vast—huge

voyage—trip

watertight—sealed so that water can't get in